YOUR KNOWLEDGE HAS VALUE

AF154610

- We will publish your bachelor's and master's thesis, essays and papers

- Your own eBook and book - sold worldwide in all relevant shops

- Earn money with each sale

Upload your text at www.GRIN.com and publish for free

Electricity and Energy Problems with Pakistan. Causes, Consequences and Sustainable Solutions

Tashif Ahmad

Bibliographic information published by the German National Library:

The German National Library lists this publication in the National Bibliography; detailed bibliographic data are available on the Internet at http://dnb.dnb.de.

ISBN: 9783656670049
This book is also available as an ebook.

© GRIN Publishing GmbH
Trappentreustraße 1
80339 München

Print and binding: Books on Demand GmbH, Norderstedt, Germany
Printed on acid-free paper from responsible sources.

The present work has been carefully prepared. Nevertheless, authors and publishers do not incur liability for the correctness of information, notes, links and advice as well as any printing errors.

GRIN web shop: https://www.grin.com/document/274407

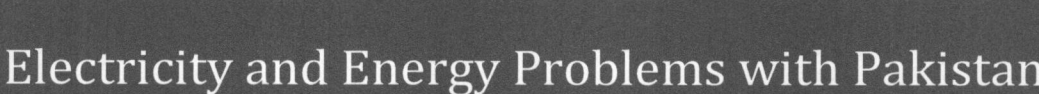

Electricity and Energy Problems with Pakistan

Causes, Consequences and Sustainable Solutions

Tashif Ahmad

Dated: 20-06-2013

Table of Contents

1. Introduction 2

2. Pakistan's Energy Sector 3

   2.1   Energy Demand and Supply 3

3. Pakistan Energy Resources 4

   3.1   Non- Renewable Energy Resources 4

   3.2   Renewable Energy Resources 4

4. Causes: How Did We Get Here 5

5. Consequences of the Energy Crisis 7

6. Sustainable Solutions to the Energy Crisis 9

7. Long Term Measures(Shifting to Altenate Sources) 9

8. Short Term Measures 13

9. Bibliography 16

10. A Report by Express Tribune 17

Introduction

Energy is considered to be the life line of an economy. It is a most vital instrument of the socio-economic development of a country. Energy is a very important factor in the production process. Energy is pivotal in running machinery in factories and industrial units, for lighting our cities and powering our vehicles etc.

There has been enormous increase in the demand of energy due to the massive industrialization and rapid population growth in comparison to the enhancement in the supply of energy production. Supply of energy is, therefore, far less than the actual demand, resultantly crisis has emerged. An energy crisis can be defined as any great bottleneck (or price rise) in the supply of energy resources to an economy. With the evolution of civilizations, the human demand for energy has continuously increased. At present , the key factor which drives the growth in energy demand include   increasing human population, modernization and urbanization .According to the united nations , the world population 6.5 billion in 2005 is to grow to 9.1 billion by 2050 and most of the population growth is expected to place in the developing world Asia and Africa.(Dinner, 1999). Poverty, hunger, disease, illiteracy and environmental degradation are the most important challenges faced by the world. Poor and inadequate access to secure an affordable means of energy in one of the crucial factors behind these issues. Electricity for example is vital for providing basic social services such as education and health, water supply and purification, sanitation, and refrigeration of essential medicines. Electricity is of course, very helpful in supporting a wide range of income generation opportunities.

The leading countries in the world in terms of population without access to electricity include India, Bangladesh, Indonesia, Nigeria, Pakistan, Congo, Ethiopia, Myanmar, Tanzania, and Kenya. With the growing world population and people's aspiration for improved life a central and collective global issue in the new century is to sustain socio-economic growth within the constraints of the earth's limited natural resource along with preserving the environment.

Pakistan's Energy Sector

Pakistan's energy infrastructure is not well developed, rather it is considered to be underdeveloped and poorly managed. Currently the country is facing severe energy crisis. Despite of strong economic growth and rising energy demand during past decade, no serious efforts have been made to install new capacity of generation. Moreover, rapid demand growth, transmission losses due to out-dated infrastructure, power theft, and seasonal reductions in the availability of hydropower have worsened the situation. Consequently, the demand exceeds supply and hence load-shedding is a common phenomenon through power shutdown.

Energy Supply and Demand:

For years, the matter of balancing Pakistan's supply against the demand for electricity has remained a largely unresolved matter. Pakistan faces a significant challenge in revamping its network responsible for the supply of electricity

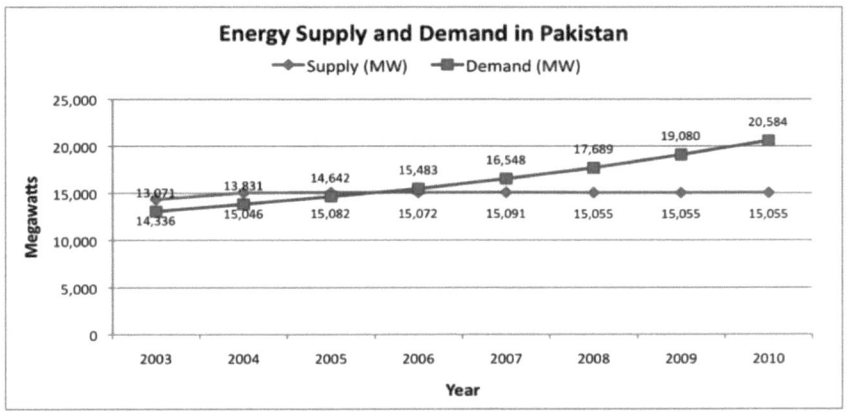

Source:http://www.pakonomy.com

At present, the power supply across the country is 9,500 MW with demand around 15,000 MW, causing the 5300 MW shortfall approximately.

Sources of Energy in Pakistan

Non-renewable resources (Fossil fuels): [Limited – Expensive]

Non-renewable resources are primarily fossil fuels emanating from remains/decomposition of animals and plants deposited deep into the earth crust and converted into oil and gas. These resources cannot be replenished. There are three main types of fossil fuels: coal, petroleum, natural gas and liquefied petroleum gas (LPG).

- ➢ Petroleum Products
- ➢ Natural Gas
- ➢ Coal

Renewable Energy Resources: (Unlimited – sustainable – clean)

Renewable energy resources are those, which are naturally replenished and come from resources such as water, sunlight, wind, rain, tides, and geothermal heat;

Hydro-Power:

Hydro power is generated by using electricity generators to extract energy from moving water. Pakistan is having rich resource of energy in hydal power; however, only 34 % of total electricity generation is coming from hydro power. Currently we are having 6555 MW against the potential of 41000 to 45000 MW.

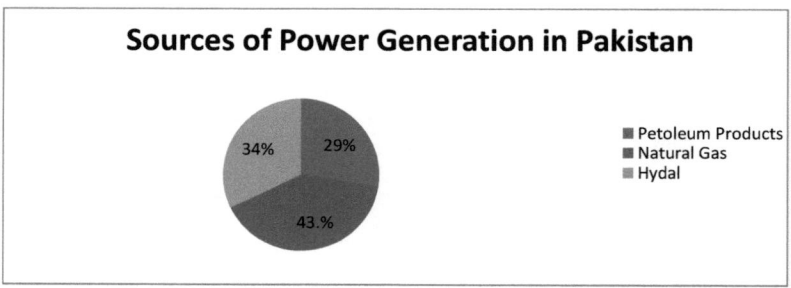

Causes: How Did We Get Here

Pakistan's power disaster traces its origins to pursuing different reasons:

Growing Power Demand:

Over the years there is larger demand of power because of;

- ➢ Increase in the Population
- ➢ Enhancement in the Lifestyle
- ➢ Industrial and Agricultural Growth
- ➢ Greater Transportation Needs

Lack of Proactive and Integrated Planning for Production of Energy:

Pakistan has had wider potentials to tap power, though, due to lack of each integrated/proactive arranging, extremely less number of manipulation producing plant were installed to encounter innovative demands. Resultantly, above the years, the gap amid power demand and supply drastically produced and nowadays opposing demand of 20000 MW, we are possessing concerning 11500 MW.

Imbalanced Energy Mix:

Energy blend in Pakistan is quite imbalance in comparison to other countries, alongside larger reliance on non-renewable resources of gas (43.7 %) and oil (29 % - bulk of that is imported). Benefits of petroleum products/crude oil fluctuate and in present Afro-Arab governmental disaster, the oil benefits are probable to raise manifold altering oil benefits in Pakistan. A rational power blend arranging must to be industrialized providing larger dependency to renewable (hydel power), original (coal) and alternative power resources (wind and solar energy).

Non-utilization of enormous indigenous energy resources:

Thar Coal:

Pakistan possesses one of the biggest coal fields in Thar, possessing reserves of extra than 175 billion tones that exceeds equivalent oil reserves of Saudi Arabia, Iran etc. In supplement to manipulation creation, this coal can be utilized for chemical and fertilizer production. Moreover, occupation endowed to workforce can be instrumental in rising GDP and commercial prosperity to countless families.

Hydal Power Generation:

Pakistan has possible of hydro resources to produce 41000 to 45000 MW, though; merely 6555 MW is presently being generated by this vital renewable resource. Four colossal hydro manipulation dams namely Kalabagh 3600 MW, Bhasha 4500 MW, Bunji 5400 MW and Dasu 3800 MW can be crafted to produce hydroelectricity. Similarly, countless tiny to medium hydro plants can be installed on streams and waterways etc.

Mopolisation:

In Pakistan, the government has monopolized the generation, transmission and distribution of electricity. This monopolization has destroyed the interests of an elite class and neglecting the masses. Tenenbaum et al. (1992) argue that "when the state owns, nobody owns; and when nobody owns, nobody cares". This then is the situation in Pakistan. Spirit of entrepreneurship. For example KESC's even after privatization is a monopoly although it may have changed hands. The monopoly does not have to worry about the competition. It can therefore continue to operate with neglect and inefficiency. Since the market is guaranteed for the monopoly, it does not have to worry about its customers or their problems. It has no incentives to respond to them and their problems. And since it has no one to answer to, it works with complete impunity.

Electricity Theft:

The problem that ails the country's power sector today is a remarkably simple one: Pakistan has an energy crisis almost entirely because a large proportion of Pakistanis are thieves who do not want to pay for electricity.

Corruption, Inefficiency, Mismanagement:

There is massive corruption in the government departments, inefficient government officials and mismanagement in handling the issue are the root cause of the current energy crisis. We have installed much greater capacity to produce electricity than our demand but it is the inefficiency and mismanagement that lead us to the current situation.

Consequences of the Energy Crisis

Economic Factors:

Energy is pivotal for running all supplementary resources and disaster of power directly influences all supplementary sectors of the economy. The economic progress is hindered by decline in agricultural productivity as well as by halting in procedures of industries. One vital factor of lower GDP and inflation of commodity prices in present years is attributed to deficits in power supply.

Agriculture Sector:

Agricultural productivity of Pakistan is decreasing due to provision of energy for running tube wells, agricultural machinery and production of fertilizers and pesticides. Thus higher energy means higher agricultural productivity.

Industrial Sector:

Nearly all Industrial units are run with the energy and breakage in energy supply is having dire consequences on industrial growth. As a result of decline in energy supply, industrial units are not only being opened, but also the existing industrial units are gradually closing.

Unemployment:

By closure of industrial units and less agricultural productivity, new employment opportunities ceased to exist and already employed manpower is shredded by the employers to increase their profit ratios. Thus energy crisis contributes towards unemployment.

Social Issues:

This factor is primarily related to the domestic usage of energy (cooking, heating and water provision). Load shedding cause unrest and frustration amongst the people and results in agitation against the government.

Poverty:

Decline in economic growth, lower agricultural productivity, unemployment and shackling industrial growth result in increasing poverty. Currently, around forty percent of our population is living beyond poverty line and this ratio is increasing day by day. Ample control of energy crisis will surely yield in curbing the menace of poverty.

Sustainable Solutions to the Crisis

As discussed above that Pakistan produces the major portion of electricity from non-renewable energy resources which are costly and less abundant whereas the renewable sources of energy are much cheaper, plentiful, environment friendly and sustainable so if adopt renewable energy sources to produce electricity we are more likely to be out of the said crisis. Furthermore we have to shift on alternate energy resources such as solar energy, wind energy, coal energy, solid waste and nuclear power to combat our current energy disaster. Here are some of the sustainable solutions;

Long term Measures :(Shifting to Alternate Energy Sources)

Wind Energy:

Wind power harnesses the power of the wind to propel the blades of wind turbines. These turbines cause the rotation of magnets, which creates electricity. Though Pakistan has potentials of wind energy ranging from 10000 MW to 50000 MW, yet power generation through wind is in initial stages in Pakistan and currently 06 MW has been installed in first phase in Jhampir through a Turkish company and 50 MW will be installed shortly. More wind power plants will be built in Jhampir, Gharo, Keti Bandar and Bin Qasim Karachi.

Solar Power:

Solar power involves using solar cells to convert sunlight into electricity, using sunlight hitting solar thermal panels to convert sunlight to heat water or air. Pakistan has potential of more than 100,000 MW from solar energy. Building of solar power plants is underway in Kashmir, Punjab, Sindh and Baluchistan. However, private vendors are importing panels / solar water heaters for consumption in the market. Alternative Energy Development Board (AEDB) is working for 20,000 solar water heaters in Gilgit Baltistan. Mobile companies have been asked by the government to shift supply of energy to their transmission towers from petroleum to solar energy panels.

Agricultural Biomass/Biodiesel:

Biomass production involves using garbage or other renewable resources such as sugarcane, corn or other vegetation to generate electricity. When garbage decomposes, methane is produced and captured in pipes and later burned to produce electricity. Vegetation and wood can be burned directly to generate energy, like fossil fuels, or processed to form alcohols. Brazil has one of the largest renewable energy programs from biomass/biodiesel in the world, followed by USA. Alternative Energy Development Board (AEDB) of Pakistan has planned to generate 10 MW of electricity from municipal waste in Karachi followed by similar projects in twenty cities of country.

Tidal Wave:

Tidal power can be extracted from Moon-gravity-powered tides by locating a water turbine in a tidal current. The turbine can turn an electrical generator, or a gas compressor, that can then store energy until needed. Coastal tides are a source of clean, free, renewable, and sustainable energy. Plans are underway in Pakistan to harness tidal energy; however, no implementation has been made so far.

Nuclear Energy:

Nuclear power stations use nuclear fission reaction to generate energy by the reaction of uranium inside a nuclear reactor. Pakistan has a small nuclear power program, with 425 MW capacity, but there are plans to increase this capacity substantially. Since Pakistan is outside the Nuclear Non-Proliferation Treaty, it is excluded from trade in nuclear plant or materials, which hinders its development of civil nuclear energy. Remaining issues in development of nuclear energy are enrichment of uranium from U235 to U238, controlling chain reaction and dumping of solid waste.

Pakistan's Nuclear Power Reactors

Reactor	Type	MWe net	Construction start	Commercial operation	Planned close
Karachi 1	PHWR	125	1966	12/72	2019
Chashma 1	PWR	300		June 2000	2040
Chashma 2	PWR	300	2005	May 2011	2051

http://www.world-nuclear.org/info/Country-Profiles/Countries-O-S/Pakistan/#.Uc-p0_ml73M

Hydrogen:

The vision of building an energy infrastructure that uses hydrogen as an energy carrier, a concept, called the hydrogen economy, consist of an economic system in which energy is supplied by renewable resources and hydrogen is used as the energy vector, medium of energy storage and transportation.

Hydrogen is one of the plentiful elements in the Universe. Chemically bound hydrogen is found abundantly on earth in water, fossil fuels and living things. Generally, it is extracted from water or from hydrocarbons. Hydrogen can also be produced through reformation of natural gas, electrolysis of water or partial oxidation of heavy fossil fuels such as diesel. The potential use of hydrogen in fuel and energy applications includes powering vehicles, running turbines or fuel cells to produce electricity and generating heat and electricity for buildings.

Shale Gas:

Shale gas is a form of natural gas that is obtained from sedimentary rock through a process called hydraulic fracturing. The US is witnessing a boom in shale gas production, and Pakistan too has substantial proven shale gas deposits. The technology is field-tested and easily acquired; there is no reason to delay shale gas exploration.

Development of the National Energy Base of Pakistan:

Pakistan is one of the most populous, geographically and strategically important countries situated in South Asia. The energy supply base of Pakistan consists of two major segments i.e. commercial and non-commercial. At the time of independence in

1947, the proportion of energy received through Commercial channels is reported to be equivalent to about 1.2 million tone oil equivalent .For the total population of about 33 million, the installed electricity generating capacity was 50 MW. (WAPDA, 1995.) The major energy consuming sectors such as, industrial, transport domestic agriculture and commercial had very little reliance on commodity. Particularly, the industrial sector was almost non-existent and motorized traveling was not very common. In the absence of a national grid, out of over 200000 towns and villages only 640 making up 7% of the population were connected to local grid. Similarly, the agriculture sector was not introduced to machinery .Thus it took almost a decade to grow the commercial energy base to take off. (Asif, 2009).

Currently, hardly 60% and 20% of the total number of households are connected respectively to the national electric grid and gas pipelines implies that the non-commercial base still makes up a considerably large proportion of the total supplies in the country. Firewood is vastly used for cooking and heating purposes. Similarly, the agriculture and livestock sectors, by producing abundant amounts of biomass in the form of crops residues and dung is collected and used as unprocessed fuel for cooking and house hold heating. According to an estimate, the biomass based non-commercial form of energy makes up almost 35% of the total consumption in the country. (Asif, 2009)

Smart Metering:

Install smart meters to curb electricity theft. Losses due to outright theft and unpaid bills approach nearly half of all electricity generated in Pakistan. What business can ever run like this?

At their current price of Rs10, 000 apiece, smart meters can return the investment made in them in about a year. Sweden, which has very minor electricity theft as compared to Pakistan, gains a third of a percent in annual GDP from smart meters. On the other hand, the energy crisis costs Pakistan about four percent of its GDP – smart meters alone can make a major difference.

Slum dwellers steal power through illegal connections, but they are less culpable than the bourgeoisie who payoff meter readers to record lower readings or refuse to pay bills altogether. Naples, Italy's third-largest city with a population of around a million, had a similar problem, but smart meters have been effective in curbing theft and tampering with meters and supply lines.

Short Term Measures

Energy Management/ Conservation:

It is said that a watt saved is a watt generated, better it is much cheaper to save than generate a watt. Therefore it is highly logical and need of the day to give prime importance to energy management / conservation.

The energy management / conservation encompasses in it ways and means of optimizing the use of energy. It should not be associated with rationing or curtailment of energy supply services. Actually it means identifying areas of wasteful use of energy and taking action to reduce the waste to a bare minimum or eliminate the waste completely. Management and technical techniques be applied to have optimum and efficient use of energy.

Settlement of the Circular Debt:

Pakistan has a circular debt in the electricity generation and distribution chain worth $5 billion, simply because the National Electric Power Regulatory Authority has proven ineffective in monitoring generators and utilities. In the West, regulatory commissions are led by energy specialists. Fixed tenures, autonomy from political interference, as well as empowerment to punish generation and distribution companies minimise chances of gaming of the system.

A strong regulator checks and balances; a weak one allows governments to run rampant. They then set tariffs as they see fit, send linemen to collect kickbacks from electricity thieves and give away electricity freebies to vote banks. Utilities go bankrupt, their slate is wiped clean, the debt re-emerges, and round and round we go.

Improve the Power distribution System:

Decouple agricultural and consumer supply networks in rural areas from each other, otherwise households benefit unfairly from the less-expensive agricultural tariffs. Gujarat in India has bifurcated the two networks to increase electricity revenues substantively. The model should be easily replicable in Pakistan.

Reducing unnecessary energy use:

> Usage of electricity saving devices.
> Awareness campaign for energy saving.
> Reduction in unnecessary transportations by developing good public transport systems and strengthening Pakistan railways.
> Reduction in industrial uses with installation of effective equipment/ energy efficient and with increasing efficiency of workforce (cost effective).
> Decreasing reliance on rental power projects, because instead of doing any good, they are increasing prices of electricity.
> Decreasing line losses by using efficient power transmission cables.

Developing new energy resources

> Tapping indigenous resources (Thar coal)
> Using renewable resources (water) by constructing new dams and hydro power plants.
> Import of natural gas by IPI (Iran Pakistan India) and TAPI (Turkmenistan, Afghanistan, Pakistan and India) pipelines.
> Import of electricity from Tajikistan -through Pak Afghan Tajikistan transmission- and Iran (approximately 1000 MW from each of them) pipelines.

Foreign Investment Should Be Facilitated in Energy Sector:

Government should facilitate investment in establishing power plants, he said, adding that production and import of natural gas be given freehand. Such initiatives will ultimately cure prevailing ills of energy sector in next 3 to 5 years once for all; foreign investors should be given incentives so that they can operate in Pakistan.

Change in lifestyles:

The nation has to draw a clear line between necessities (lighting, fans, TVs, computers, etc) and luxuries (air conditioners, microwaves, etc). There is not enough electricity to meet both requirements. We should utilize daylight as possible, and our government has already taken steps about it. An early start and early end is recommended rather than having opening hours from afternoon until late at night. Air-conditioning, usually a sign of a luxurious lifestyle, needs to be ped.

References

. Asif, M. (2011). *Energy Crisis in Pakistan Origin Challanges and Sustainable Solutions.* Karachi: Oxford University Press.

. *Malik, A. (2012). Power crisis in Pakistan: A crisis in governance? (PIDE Monograph Series). Islamabad, Pakistan: Pakistan Institute of Development Economics.*

. *Government of Pakistan (2010). Pakistan Economic Survey 2008-09*

http://www.pakistantoday.com.pk/2013/05/18/news/national/power-crisis-worsens-as-shortfall-hits-5300-mw/#sthash.plleIwYx.dpuf

http://tribune.com.pk/story/216609/consequences-of-energy-crisis-as-export-orders-become-harder-to-meet-textile-sector-turns-to-local-market/

http://www.world-nuclear.org/info/Country-Profiles/Countries-O-S/Pakistan/#.Uc-p0_ml73M

http://tribune.com.pk/story/561071/the-game-plan-a-few-pointers-for-solving-pakistans-energy-crisis/

http://www.jstor.org/stable/41261233

http://www.jstor.org/stable/41258572

http://www.jstor.org/stable/41260897

Report

On

Why power outages are mostly our own fault?
By
Express Tribune

For all the enormous technical complications involved in running a nationwide electricity grid, the problem that ails the country's power sector today is a remarkably simple one: Pakistan has an energy crisis almost entirely because a large proportion of Pakistanis are thieves who do not want to pay for electricity.

This week, *The Express Tribune* attempts to take a comprehensive look at the power sector and examine the causes for the massive power shutdowns. Our goal in this report is not to look at what the problem is currently, but how it arose in the first place and what can be done to fix it.

The picture we come up with is grim. At virtually every stage of the energy cycle, there are massive problems with either the government's planning or execution, and frequently both, a fact that becomes easier to understand when one realises that the government has never had a comprehensive national energy policy.

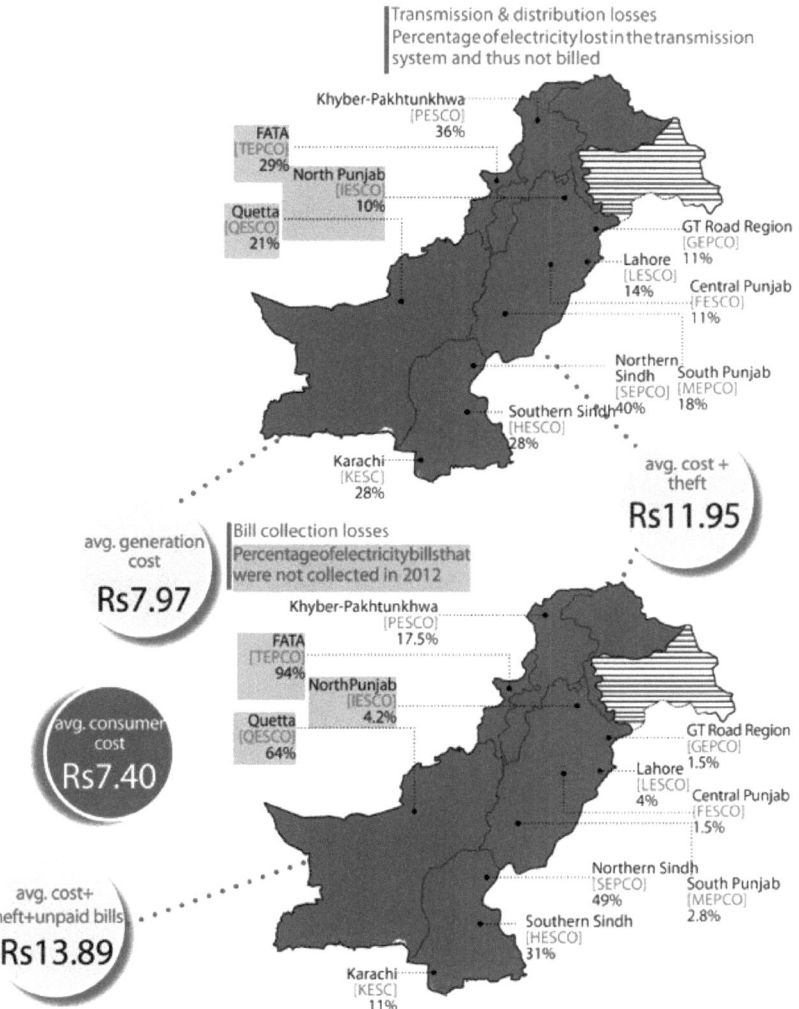

Transmission & distribution losses
Percentage of electricity lost in the transmission system and thus not billed

Khyber-Pakhtunkhwa
[PESCO]
36%

FATA
[TEPCO]
29%

North Punjab
[IESCO]
10%

Quetta
[QESCO]
21%

GT Road Region
[GEPCO]

Lahore 11%
[LESCO]
14%

Central Punjab
[FESCO]
11%

Northern
Sindh South Punjab
[SEPCO] [MEPCO]
Southern Sindh 40% 18%
[HESCO]
28%

Karachi
[KESC]
28%

avg. cost +
theft

Rs11.95

avg. generation
cost

Rs7.97

Bill collection losses
Percentage of electricity bills that were not collected in 2012

Khyber-Pakhtunkhwa
[PESCO]
17.5%

FATA
[TEPCO]
94%

North Punjab
[IESCO]
4.2%

Quetta
[QESCO]
64%

GT Road Region
[GEPCO]

Lahore 1.5%
[LESCO]
4%

Central Punjab
[FESCO]
1.5%

avg. consumer
cost

Rs7.40

Northern Sindh
[SEPCO]
49%

South Punjab
[MEPCO]
2.8%

Southern Sindh
[HESCO]
31%

avg. cost+
theft+unpaid bills

Rs13.89

Karachi
[KESC]
11%

Our special report is divided up into four segments, each one covering a different segment of the energy chain: sources of primary energy, power generation, transmission, and distribution.

The problem starts at the very beginning, with a fuel mix that is skewed towards the most expensive kinds. There is an absurdly large reliance on oil, for instance, which is virtually unheard of in countries without substantial oil reserves of their own. Not nearly enough investment has gone into increasing hydroelectric power generation, which is the most expensive to set up but the cheapest to run on a per unit basis.

At every subsequent stage, the problem keeps on compounding, the mistakes of the previous stage being magnified by the inefficiencies of every successive segment of the supply chain.

The poor fuel mix translates into electricity generation being more expensive than it needs to be. The expensive electricity, coupled with exceedingly poor rule of law, in turn encourages outright theft of electricity and leaves little for maintenance of the infrastructure, which causes yet more rises in the cost of producing electricity. And finally, an unwillingness of the government to force people to pay their bills results in yet more losses.

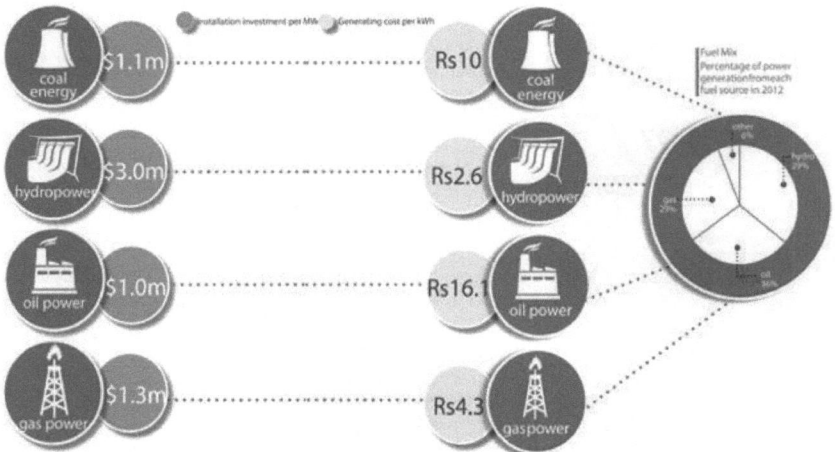

The key problem in all of this, however, is the propensity of a large proportion of Pakistanis to steal electricity and the government's unwillingness to confront them. Our analysis indicates that if transmission losses in Pakistan were kept at a level similar to that of the United States, and if bill collection were close to 100%, tariffs would need to rise by a relatively meagre 7.7% from their current average levels.

The problem, of course, is that theft is not going away in Pakistan any time soon. And so we have, effectively, a nation that steals electricity, pays no taxes, and then complains when the government runs out of money to pay for the electricity that they stole in the first place. In the power sector, the debt is very clearly not the only thing that is circular.

The problem cannot be resolved until the stealing is stopped. Locking up everyone who steals electricity is clearly not possible, but incentivising good behaviour is very much within reach. In this, the government would do well to carefully examine the model of the Karachi Electric Supply Company, which has spent considerable time and resources in figuring out which areas have high theft and which have low theft, and then provided relatively uninterrupted supply to areas where people do not steal electricity.

In the weeks since the election, the country is going through a unique moment: when two of the largest political parties in the country broadly agree on the contours of a solution to the energy crisis. This is a moment that is unlikely to return in the future. If the country's leaders fail now, it may be lights out on the Pakistani economy for a very long time to come.

Published in The Express Tribune, June 3rd, 2013.

YOUR KNOWLEDGE HAS VALUE

- We will publish your bachelor's and master's thesis, essays and papers

- Your own eBook and book - sold worldwide in all relevant shops

- Earn money with each sale

Upload your text at www.GRIN.com and publish for free